Grade 3

Discovery EDUCATION™ | SCIENCE TECHBOOK

California
Unit 2
Life Cycles for Survival

To obtain permission(s) or for inquiries, submit a request to:

Discovery Education, Inc.
4350 Congress Street, Suite 700
Charlotte, NC 28209
800-323-9084
Education_Info@DiscoveryEd.com

ISBN 13: 978-1-68220-542-6

Printed in the United States of America.

3 4 5 6 7 8 9 10 CWM 27 26 25 24 23 22   B

Acknowledgments

Acknowledgment is given to photographers, artists, and agents for permission to feature their copyrighted material.

Cover and inside cover art: Parilov / Shutterstock.com

Table of Contents

Unit 2: Life Cycles for Survival

Letter to the Parent/Guardian vi

Unit Overview vii

Anchor Phenomenon: Honeybees on a Comb 2

Unit Project Preview: Honeybee Population Loss 4

Concept 2.1 Life Cycles

Concept Overview 6

Wonder 8

Investigative Phenomenon: Planting Seeds 10

Learn 20

Share 40

Concept 2.2 Inherited Traits

Concept Overview 50

Wonder 52

Investigative Phenomenon: Parents and Their Young 54

Learn 60

Share 82

Concept 2.3 Working in Groups

Concept Overview 90

   Wonder 92

      Investigative Phenomenon: Whales in Groups 94

   Learn 100

   Share 114

Unit Wrap-Up

Unit Project: Honeybee Population Loss 124

Grade 3 Resources

Bubble Map R3

Safety in the Science Classroom R4

Vocabulary Flash Cards R7

Glossary R21

Index R46

Discovery
EDUCATION

Dear Parent/Guardian,

This year, your student will be using Science Techbook™, a comprehensive science program developed by the educators and designers at Discovery Education and written to the California Next Generation Science Standards (NGSS). The California NGSS expect students to act and think like scientists and engineers, to ask questions about the world around them, and to solve real-world problems through the application of critical thinking across the domains of science (Life Science, Earth and Space Science, Physical Science).

Science Techbook is an innovative program that helps your student master key scientific concepts. Students engage with interactive science materials to analyze and interpret data, think critically, solve problems, and make connections across science disciplines. Science Techbook includes dynamic content, videos, digital tools, Hands-On Activities and labs, and game-like activities that inspire and motivate scientific learning and curiosity.

You and your child can access the resource by signing in to www.discoveryeducation.com. You can view your child's progress in the course by selecting the Assignment button.

DISCOVERY
EDUCATION

Science Techbook is divided into units, and each unit is divided into concepts. Each concept has three sections: Wonder, Learn, and Share.

Units and Concepts Students begin to consider the connections across fields of science to understand, analyze, and describe real-world phenomena.

Wonder Students activate their prior knowledge of a concept's essential ideas and begin making connections to a real-world phenomenon and the **Can You Explain?** question.

Learn Students dive deeper into how real-world science phenomenon works through critical reading of the Core Interactive Text. Students also build their learning through Hands-On Activities and interactives focused on the learning goals.

Share Students share their learning with their teacher and classmates using evidence they have gathered and analyzed during Learn. Students connect their learning with STEM careers and problem-solving skills.

Within this Student Edition, you'll find QR codes and quick codes that take you and your student to a corresponding section of Science Techbook online. To use the QR codes, you'll need to download a free QR reader. Readers are available for phones, tablets, laptops, desktops, and other devices. Most use the device's camera, but there are some that scan documents that are on your screen.

For resources in California Science Techbook, you'll need to sign in with your student's username and password the first time you access a QR code. After that, you won't need to sign in again, unless you log out or remain inactive for too long.

We encourage you to support your student in using the print and online interactive materials in Science Techbook, on any device. Together, may you and your student enjoy a fantastic year of science!

Sincerely,

The Discovery Education Science Team

Unit 2
Life Cycles for Survival

Honeybees on a Comb

Quick Code:
ca3351s

Have you ever been stung by a bee? You were probably angry afterward, but please don't hurt the bees. Honeybees work together to pollinate flowers for our fruits and vegetables. They also use nectar from the plants to make honey. In this unit, you will explore how animals grow, why parents are important, and how some animals (including honeybees) work together.

Honeybees on a Comb

Discovery EDUCATION

Think About It

Look at the photograph. **Think** about the following questions.

- How can the four stages of a life cycle be used to predict what will happen next?

- What do parents and their offspring have in common?

- How do inherited traits help offspring survive?

- Why do animals form groups?

Honeybees

Solve Problems Like a Scientist

Quick Code:
ca3352s

Unit Project:
Honeybee Population Loss

In this project, you will analyze data about the loss of honeybee colonies. Then, you will research methods to reduce bee deaths and analyze how effective these methods are.

Bees on a Honeycomb

SEP **Asking Questions and Defining Problems**

Ask Questions About the Problem

You are going to research methods to reduce honeybee loss. Then, you will evaluate how effective these methods are using what you know about animal life cycles and working in groups. **Write** some questions you can ask to learn more about the problem of honeybee colony loss. As you learn about animal life cycles and how some animals work in groups, **write** down the answers to your questions.

Life Cycles

Student Objectives

By the end of this lesson:

☐ I can develop models to compare patterns in the birth, growth, reproduction, and death of various plants and animals.

☐ I can develop models to describe how animals change throughout their life cycles.

Key Vocabulary

☐ cycle
☐ endangered
☐ extinct
☐ germination
☐ life cycle

☐ lifespan
☐ mature
☐ metamorphosis
☐ offspring
☐ organism

☐ seed
☐ seedling

Quick Code:
ca3353s

Can You Explain?

How are life cycles of various organisms the same and different?

_____

_____

_____

_____

_____

_____

_____

_____

_____

Quick Code:
ca3355s

Activity 2
Ask Questions Like a Scientist

Quick Code:
ca3356s

Planting Seeds

Look at the picture. Then, **answer** the questions.

Let's Investigate Planting Seeds

SEP Constructing Explanations and Designing Solutions

CCC Stability and Change

What do you think will happen after this student plants the seeds?

_____

_____

_____

Create a chart to describe how the seed will change over time.

Before:	After:

Changes:

Analyze Like a Scientist

Plants and Animals

Quick Code:
ca3357s

Read the text. As you read, **highlight** information you can use as evidence to support your ideas to answer the Can You Explain? question. Then, **complete** the activities that follow.

Plants and Animals

What do lion cubs and a Joshua tree have in common?

Lion Cubs

Joshua Tree

SEP **Obtaining, Evaluating, and Communicating Information**

They both start as smaller **organisms** and grow into larger organisms. A Joshua tree can live to be 150 years old.

A lion in the wild can live up to 14 years. Why is a lion's **lifespan** so much shorter than that of a Joshua tree?

Mayfly

Throughout this concept, you will learn about how different organisms change over time. There are some organisms, like a California oak tree, that may live for 300 years, while a mayfly lives only 24 hours. As you work through the concept, think about patterns that exist across all organisms.

California Black Oak

Compare the two organisms on the Venn Diagram.
Write characteristics of the lion cubs in one circle,
characteristics of the Joshua Tree in the other, and
characteristics they share in the middle.

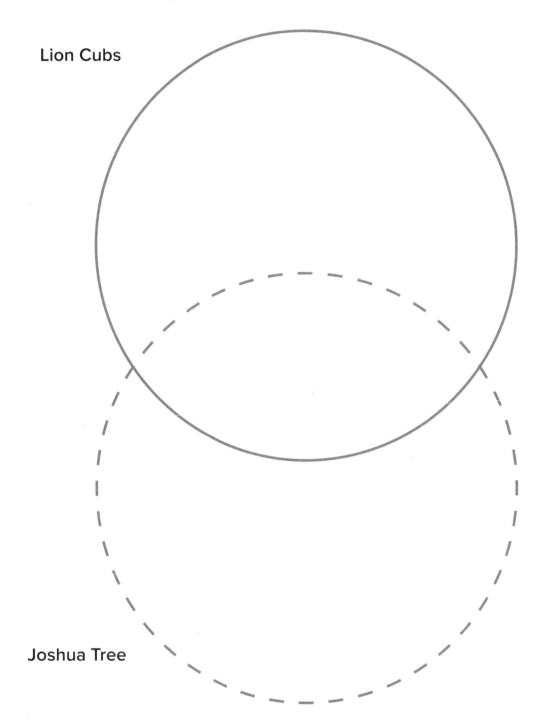

Lion Cubs

Joshua Tree

Now, **choose** either the lion cub or the Joshua tree.

Draw how the organism will appear two years later.

Activity 4
Observe Like a Scientist

Mayfly

Look at the picture. **Think** about the life of a mayfly. Then, **talk** about the questions.

Mayfly

Talk Together

How does the mayfly change in its 24 hours of life? What does the mayfly do in the 24 hours? How does the mayfly make sure more mayflies are born?

Activity 5

Evaluate Like a Scientist

Quick Code:
ca3359s

What Do You Already Know About Life Cycles?

Nonliving or Dead

What is the difference between something that is nonliving and something that is dead? **Write** your ideas.

_____

_____

_____

_____

_____

_____

_____

_____

_____

_____

Life Cycle Terms

Match each part of the animal life cycle to its description.

Animals use these to produce offspring; they may be external to the animal, in a hard shell, or internal.

adult

The end of an animal's life.

egg

The fully formed individual organism; usually has the ability to reproduce.

metamorphosis

A complete change in body form that some animals go through in their life cycle.

death

Discovery EDUCATION

How Do Seeds Become Seedlings?

How does a seedling grow from such a small seed? **Write** your ideas.

_____

_____

_____

_____

_____

What Are the Stages of the Life Cycle of a Plant?

Activity 6

Investigate Like a Scientist

Quick Code:
ca3360s

Hands-On Investigation: From a Seed to a Plant

In this investigation, you will observe the different stages of lima bean development. You will take measurements to document how the lima bean grows and changes.

Make a Prediction

Develop a hypothesis about how the lima bean will develop.

_____

_____

_____

_____

_____

SEP	**Asking Questions and Defining Problems**
SEP	**Planning and Carrying Out Investigations**
CCC	**Patterns**

What materials do you need? (per group)

- Sprouted lima beans at various stages of development

- Lima beans (pre-soaked)

- Seeds, lima beans

- Forceps

- Toothpicks

- Hand lens

- Knife, plastic

- Metric ruler

- Paper plate

- Construction paper

- Camera

- Microscope

- Microscope slides

- Graph paper

- Paper, blank, white, 8.5" × 11"

- Markers

What Will You Do?

1. **Observe** the outside and inside of the dry and germinated lima beans.

2. **Measure** the beans and record your observations.

3. **Come back** every day and repeat your measurements.

4. **Create** a graph that organizes and displays your results.

Create a graph that shows the height of the seedling over time

> **Lima Bean**

Think About the Activity

What parts of the seed were you able to see in the activity?

_____

_____

If the seedling is so small, why is the seed so big?

_____

_____

_____

_____

If we take these seeds and plant them, what do you think will happen?

_____

_____

_____

Draw some pictures of what the seedlings look like at different times. **Label** your drawings.

_____ _____ _____

_____ _____ _____

_____ _____ _____

Activity 7

Analyze Like a Scientist

Stages of an Animal's Life Cycle

Quick Code:
ca3361s

Read the text about animal life cycles. **Highlight** patterns in the life cycles of animals and amphibians.

Stages of an Animal's Life Cycle

Video

Life Cycles of Mammals

All animals, including humans, have a **life cycle**. However, mammals, fish, birds, insects, amphibians, and reptiles have different types of life cycles.

For example, most animals lay eggs. Only some give birth to live **offspring**.

Discovery EDUCATION

The offspring of mammals stay with and are cared for by their parents. In contrast, most fish offspring must fend for themselves from the moment they are born.

Many organisms, like insects, change in appearance as they change from newborn offspring into adults. This process is called **metamorphosis**.

These changes not only affect what organisms look like, but also where they live, what they eat, and what eats them. Many types of animals undergo metamorphosis, including insects and some amphibians.

Salmon Eggs

Ladybug Metamorphosis

Video

Amphibian Life Cycles

Activity 8

Observe Like a Scientist

Insect Life Cycle: Metamorphosis

Watch the video. **Notice** how the painted lady butterfly and the milkweed bug change. Then, **draw** a timeline for each of their life cycles.

Quick Code: ca3362s

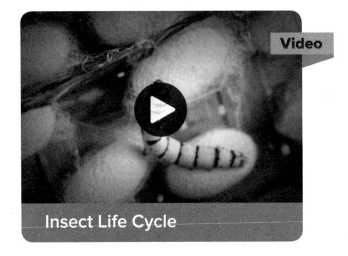

Insect Life Cycle

CCC Patterns

Painted Lady Butterfly

Milkweed Bug

Activity 9
Observe Like a Scientist

Quick Code:
ca3363s

Animals: Growing Up

Complete the Growing Up part of the interactive for an animal. **Record** your data in the data chart.

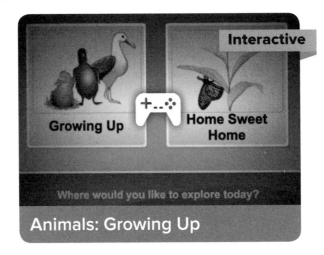

Growing Up   Home Sweet Home

Where would you like to explore today?

Animals: Growing Up

CCC **Scale, Proportion, and Quantity**

CCC **Patterns**

Name of Animal: _____

Growth Stage	Drawing	Fact
Fetal		
Infant		
Juvenile		
Adult		

Activity 10
Evaluate Like a Scientist

Quick Code:
ca3364s

Life Cycle Models

Different types of animals have different life cycles. They can undergo complete metamorphosis, incomplete metamorphosis, or no metamorphosis. **Examine** the life cycles of the various organisms shown. **Circle** the words that describe each life cycle.

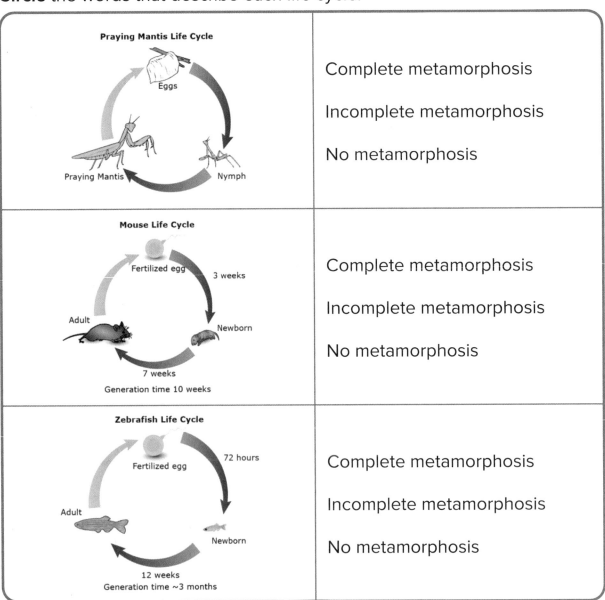

Praying Mantis Life Cycle (Eggs, Nymph, Praying Mantis)	Complete metamorphosis Incomplete metamorphosis No metamorphosis
Mouse Life Cycle (Fertilized egg, 3 weeks, Newborn, 7 weeks, Adult, Generation time 10 weeks)	Complete metamorphosis Incomplete metamorphosis No metamorphosis
Zebrafish Life Cycle (Fertilized egg, 72 hours, Newborn, 12 weeks, Adult, Generation time ~3 months)	Complete metamorphosis Incomplete metamorphosis No metamorphosis

CCC Patterns

Discovery EDUCATION

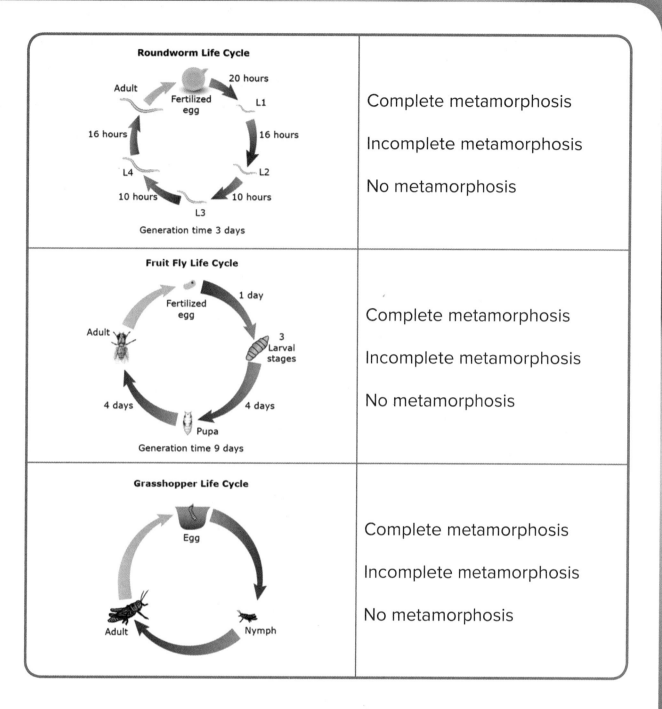

Roundworm Life Cycle	Complete metamorphosis
	Incomplete metamorphosis
	No metamorphosis

Roundworm Life Cycle

Adult — Fertilized egg — 20 hours — L1 — 16 hours — L2 — 10 hours — L3 — 10 hours — L4 — 16 hours

Generation time 3 days

Fruit Fly Life Cycle

Fertilized egg — 1 day — 3 Larval stages — 4 days — Pupa — 4 days — Adult

Generation time 9 days

Complete metamorphosis

Incomplete metamorphosis

No metamorphosis

Grasshopper Life Cycle

Egg — Nymph — Adult

Complete metamorphosis

Incomplete metamorphosis

No metamorphosis

Activity 11
Observe Like a Scientist

Life Cycle Stages

© Discovery Education | www.discoveryeducation.com ● Image: (a) David Malan / DigitalVision / Getty Images, (b) Icon made by Freepik from www.flaticon.com

Complete the seeds portion of the interactive. **Think** about how the stages of the plant life cycle are similar to those of animals. Then, **discuss** the question.

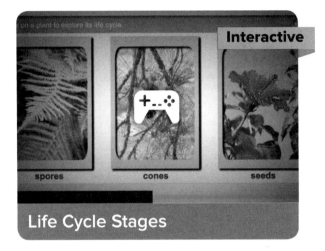

 **Talk Together**

What are the similarities and differences between an animal life cycle and the flowering plant life cycle?

CCC **Patterns**

Activity 12

Analyze Like a Scientist

Stages of a Plant Life Cycle

Quick Code:
ca3366s

Read the text about plant life cycles. **Underline** the text that corresponds to different stages of life cycles.

Stages of a Plant Life Cycle

Like all living things, plants have a life **cycle**. Most plants start life as **seeds**. In the right environment, seeds will **germinate** and grow into **seedlings**. Over time, a seedling will grow into an adult plant.

Life Cycle of a Flowering Plant

> **SEP** Developing and Using Models
> **SEP** Obtaining, Evaluating, and Communicating Information
> **CCC** Patterns

Vegetable Plant

California Redwood

Different types of plants grow at different rates. Some plants take weeks to **mature**. Others take years. The life cycle of plants can vary widely. Some plants, like the vegetables planted by many gardeners, only live for a few months. During that time, they grow from a seedling to a mature **plant** and produce vegetables before they die.

Other plants can have much longer life spans. Redwood trees, which grow in coastal regions in California and Oregon, can live for 300 years. No matter how long a plant lives, the last part of the life cycle is the same—death. Like all living things, plants eventually die.

Activity 13

Evaluate Like a Scientist

Quick Code:
ca3367s

Plant Life Cycle

Plants that reproduce with seeds have very similar life cycles. **Number** the steps in the life cycle in the correct order, starting with a seed.

A plant produces a seed. _____

The plant dies. _____

The seed germinates. _____

The plant continues to grow and become an adult plant. _____

A seedling grows. _____

CCC **Patterns**

What Patterns Can You Notice in Life Cycles?

Activity 14

Observe Like a Scientist

Quick Code:
ca3369s

Observable Patterns

Look at the life cycles shown. **Label** the time period for each stage of the life cycles. Then, **talk** about the question.

Life Cycle of Bees

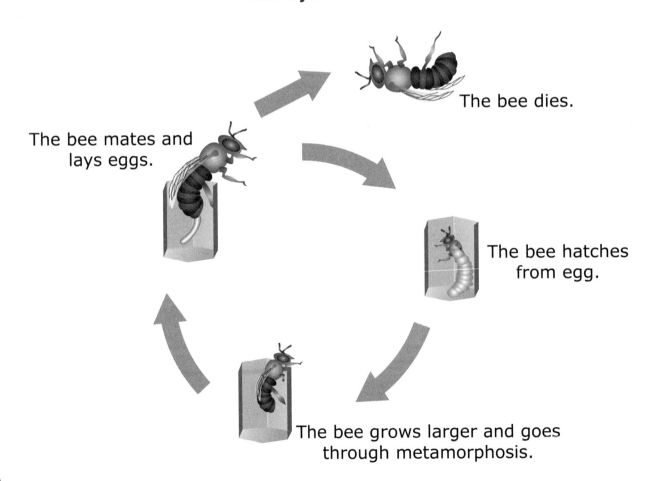

The bee dies.

The bee mates and lays eggs.

The bee hatches from egg.

The bee grows larger and goes through metamorphosis.

Life Cycle of a Pea

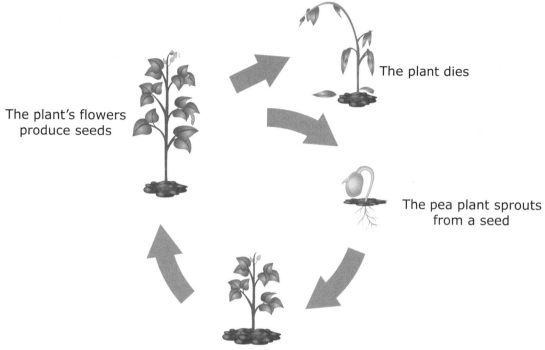

The plant dies

The plant's flowers produce seeds

The pea plant sprouts from a seed

The pea plant grows larger

 **Talk Together**

How does the time spent in each life cycle stage differ for bees and pea plants?

SEP	**Developing and Using Models**
CCC	**Patterns**
CCC	**Stability and Change**

Activity 15

Evaluate Like a Scientist

Quick Code:
ca3370s

Life Cycle of a Poppy and a Frog

Life Cycle of a Poppy

Model the life cycle of a flowering plant called the California poppy. **Number** the pictures in the correct order.

_____ _____ _____ _____

Life Cycle of a Frog

Model the life cycle of a frog. **Number** the pictures in the correct order.

_____ _____ _____ _____ _____

SEP Developing and Using Models

© Discovery Education | www.discoveryeducation.com • Image: David Malan / DigitalVision / Getty Images

Activity 16

Record Evidence Like a Scientist

Quick Code:
ca3371s

Planting Seeds

Now that you have learned about life cycles, look again at Planting Seeds. You first saw this in Wonder.

Let's Investigate Planting Seeds

Talk Together

How can you describe Planting Seeds now? How is your explanation different from before?

Look at the Can You Explain? question. You first read this question at the beginning of the lesson.

Can You Explain?

How are life cycles of various organisms the same and different?

Now, you will use your new ideas about Planting Seeds to answer a question.

Write a question. You can use the Can You Explain? question or one of your own. You can also use one of the questions that you wrote at the beginning of the lesson.

My Question

SEP **Constructing Explanations and Designing Solutions**

Now, **complete** the chart to answer your question.

Claim: _____

Reasoning	Evidence

 **in Action**

Quick Code:
ca3372s

 Activity 17

Analyze Like a Scientist

Saving Endangered Species

Read the text about endangered species. Then, **watch** the videos about endangered species. **Underline** what can cause species to become extinct.

Saving Endangered Species

Sometimes, the life cycles of animals get interrupted. Instead of dying when they are old, animals may die while they are still young. If this occurs to just a few animals in a species with a large population, it does not cause problems for the species. If many or much more than the usual number of young animals of a species die without reproducing, the population will start to decline. When this happens over time, the species can become **endangered**. If the situation does not change, the species could become **extinct**.

SEP **Obtaining, Evaluating, and Communicating Information**

Endangered Species

One of the roles of zoos around the world is to try to help protect animals that are endangered. Zoologists are key in this work. Zoologists are scientists who study animals and the ways in which animals interact with their environments. Some zoologists work in an animal's natural habitat, and others work with animals in captivity. Some zoologists do research in laboratories. Zoologists around the world work together to protect endangered species. One of the ways they do this is with managed breeding programs, such as those run by zoologists at the London Zoo.

United Kingdom: London
Zoo Census

To be able to help endangered species, zoologists must understand the animal's life cycle. They need to know how the animal grows and develops. They help provide the right conditions for growth and development. They can also make sure that conditions are right for the animals to reproduce.

Endangered Species

You have learned that some species are endangered. What criteria are used to make this determination? **Review** the information in the table. Then **answer** the question.

Classification	Population Reduction Rate
Vulnerable Species	30–50% population decline
Endangered Species	50–70% population decline
Critically Endangered Species	≥ 80–90% population decline
Extinct in the Wild	Only survives in captivity or as a population well outside its established range
Extinct	No remaining individuals of the species

The American bison population has declined in North America over the last two centuries. Historically, the population was believed to be made up of more than 30 million bison. The population dropped below 250 in 1900 because of hunting. Through conservation efforts that started in the 1920s, the American free-ranging bison population is now around 9,000. Based on its population reduction, how would the American bison be classified now?

A. vulnerable species

B. endangered species

C. critically endangered species

D. extinct in the wild

E. extinct

Activity 18

Evaluate Like a Scientist

Quick Code:
ca3373s

Review: Life Cycles

Think about what you have read and seen. What did you learn?

Write down some core ideas you have learned. **Review** your notes with a partner. Your teacher may also have you take a practice test.

_____

_____

_____

_____

_____

_____

_____

_____

_____

_____

_____

SEP   **Obtaining, Evaluating, and Communicating Information**

Talk Together

Think about the honeybee population loss you read about in Get Started. Use your new ideas about life cycles to discuss what might be causing honeybee loss.

Inherited Traits

Student Objectives

By the end of this lesson:

☐ I can provide evidence that plants inherit traits from their parents and also differ from their parents.

☐ I can reason and analyze patterns to predict the traits offspring will inherit from their parents.

☐ I can use evidence to explain how the variations in characteristics provide advantages in surviving, finding mates, and reproducing.

Key Vocabulary

☐ adaptation

☐ artificial selection

☐ camouflage

☐ characteristic

☐ generation

☐ inherit

☐ pollen

☐ trait

Quick Code: ca3374s

Activity 1

Can You Explain?

How do the traits offspring inherit help them survive?

_____

_____

_____

_____

_____

_____

_____

_____

_____

_____

Quick Code:
ca3375s

Activity 2

Ask Questions Like a Scientist

Quick Code:
ca3377s

Parents and Their Young

Look at the picture. Then, **answer** the questions.

Let's Investigate Parents and Their Young

SEP **Asking Question and Defining Problems**

What are the differences between the young and adult seal?

_____

_____

_____

_____

How can mother seals know which is their pup?

_____

_____

_____

_____

What questions do you have about parents and their young?

_____

_____

_____

_____

Activity 3

Evaluate Like a Scientist

Quick Code: ca3378s

What Do You Already Know About Inherited Traits?

Can You Determine the Parents?

Match the kitten to its parent. Then, describe which characteristics helped you match the kitten to its parent.

Parent Cats	**Kittens**

© Discovery Education | www.discoveryeducation.com ● Image: (a) KeithSzafranski / E+ / Getty Images (b) John Greim / LightRocket / Getty Images (c) Natalia Kolesnikova / AFP / Getty Images (d) Shirlaine Forrest / WireImage (e) ullstein bild / ullstein bild / Getty Images (f) DEA/D. ROBOTTI / De Agostini / Getty Images (g) Jody Trappe Photography / Moment / Getty Images

Parental Plants

Match the parent plants to their offspring plants.

Parent Plants

Offspring Plants

Adaptations and Organisms

A list of **adaptations** found in different organisms is shown.
Match each adaptation with the phrase that best describes
how the adaptation helps the organism survive.

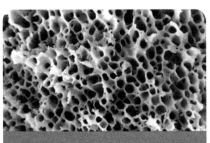

Lightweight bones

Thick layer of fat under the skin

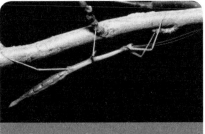

Camouflage

Long, straw-shaped tongue

Helps the organism hide from enemies

Helps the organism collect food

Helps the organism fly

Helps the organism stay warm

Identifying Adaptations

Observe the images of the plants that grow in two different environments. The plants on the left grow in a harsh, dry environment. The plants on the right grow in a mild, moist environment. Use evidence and reasoning to **compare** the adaptations of these plants. **Explain** how the adaptations help the plants survive in their environments.

_____

_____

_____

_____

_____

_____

_____

_____

What Traits Can Be Passed Down from Parents?

Activity 4

Investigate Like a Scientist

Quick Code:
ca3379s

Hands-On Investigation: Plant Generations

In this investigation, you will compare the characteristics of parent and offspring plants by observing the height of parent plants compared with the height of their offspring.

Make a Prediction

Develop a hypothesis about the relationship in heights between parent and offspring plants.

_____

_____

_____

_____

CCC **Patterns**

What materials do you need? (per group)

- Plastic cup, 9 oz
- Seeds, radish
- Soil, potting
- Water
- Metric ruler
- Lab apron (per student)
- Disposable gloves (per student)

What Will You Do?

1. **Plan** your investigation. **Write** your plan in the box, and get approval from your teacher.

2. **Plant** your first seeds. Are your plants tall or short?

_____

_____

_____

_____

_____

_____

_____

_____

_____

_____

_____

_____

_____

_____

3. Now, **pollinate** two tall plants and two short plants. Measure the offspring of the parents.

4. **Record** your observations in the data table.

Data Table for Parent and Offspring Plants:

Day Height is Measured	Parent A (Tall)		Parent B (Short)	
	Tall Parent Plant	Offspring of A	Short Parent Plant	Offspring of B
4				
8				
12				
16				
20				
24				

Think About the Activity

Following the investigation, **analyze** your data and **answer** the questions.

How were the offspring plants like the parent plants? How were they different?

Similarities	Differences

Discovery
EDUCATION

How did your data compare with that of your classmates?

_____

_____

_____

_____

_____

What pattern do you see in your observations?

_____

_____

_____

_____

_____

_____

_____

Activity 5
Analyze Like a Scientist

Quick Code:
ca3380s

Traits

Read the text about traits. **Highlight** examples of traits in the text.

Traits

Horses Grazing

Did you look in a mirror today? What color are your eyes and your hair? These physical features are called **characteristics**.

CCC **Patterns**

All organisms have characteristics, also called **traits**. For example, a horse could have a brown coat. Another horse could have a black coat. Some horses can run fast. Others are strong and can pull heavy loads. Horses may also behave differently. One horse might be friendly to people, while another horse might be shy and run away. All organisms have a variety of traits.

How Are Siblings Similar and Different?

Activity 6
Observe Like a Scientist

Quick Code:
ca3381s

Similarities of Parents and Offspring

Complete the Animal and Plant portions of the interactive.
Then, **answer** the questions.

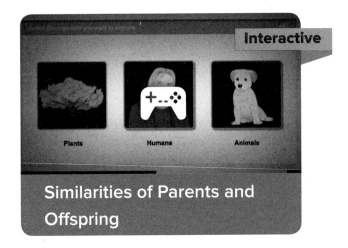

Interactive

Similarities of Parents and Offspring

| SEP | **Developing and Using Models** |
| CCC | **Patterns** |

What are inherited traits?

_____

_____

_____

_____

_____

_____

Why can't horse parents have a baby cow?

_____

_____

_____

_____

_____

_____

Siblings *Brother or sister*

Read the text about siblings. **Highlight** evidence in the text that supports or does not support the claim "All baby animals look exactly like their parents."

Siblings

Look at these adorable cats! Though they are the same type of cat and have the same parents, they have different characteristics. One has mostly white fur, and the other has mostly black and orange fur.

Calico Cat Siblings

SEP **Engaging in Argument from Evidence**

CCC **Patterns**

Offspring **inherit** traits from both of their parents. Brothers and sisters, also known as siblings, will still look similar in many ways.

The cats in the photograph have similar ears and bodies. However, each offspring has its own unique mixture of traits. One cat may have the same colors as the father, another the same colors as the mother, or they may have a mixture of both.

All siblings will look similar to their parents but not exactly like them.

Puppies

How Do Traits Passed Down from a Parent Help
a Plant or Animal Survive?

Activity 8

Evaluate Like a Scientist

Quick Code:
ca3383s

Parents and Their Offspring

Match the offspring to the parent based on their ears. Some
parents may have more than one offspring that matches
them.

CCC **Patterns**

Parents

Offspring

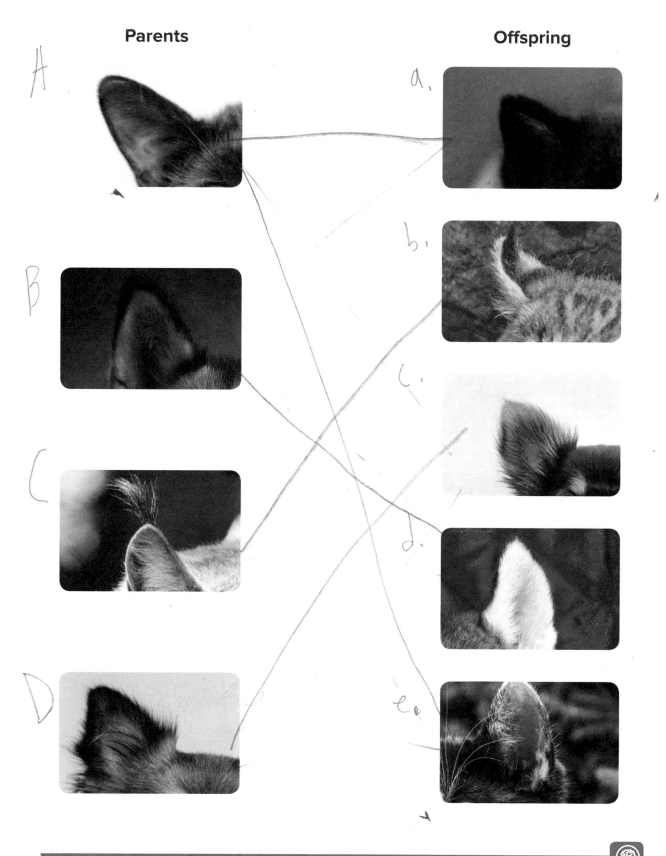

A

B

C

D

a.

b.

c.

d.

e.

Analyze Like a Scientist

Inherited Traits and Survival

Read the text about inherited traits. **Highlight** evidence in the text that describes the traits flowers have in common. **Underline** evidence in the text that describes the traits that are unique to each species.

Inherited Traits

These two flowers have some common traits and some differences. The flowers have common parts such as the roots, stems, and leaves. Their differences are more noticeable. The shape of the flower and its petals are unique to each flower.

Sunflower

Pink Rose

SEP **Engaging in Argument from Evidence**

They both have traits that help them survive. The rose has thorns to prevent predators from taking a bite, and the sunflower makes a lot of seeds to increase the chance of reproduction.

Read the following text about survival. **Label** the sentences that support each of the following claims:

Claim 1: Taller flowers have a better chance of survival.

Claim 2: Shorter flowers have a better chance of survival.

Claim 3: Larger flowers have a better chance of survival.

Survival

Individuals of the same species may look alike but they are rarely identical.

Each individual is slightly different.

These differences in characteristics may increase the chances of an individual surviving.

Look closely at the sunflowers.

Field of Sunflowers

Some of the flowers are taller than the others.

The tall flowers probably had tall parents.

Those taller flowers will have a better chance of survival than the shorter flowers because they will be better able to get sunlight.

The shorter flowers may have all of their sun blocked and may not be able to get what they need for survival.

Bigger sunflowers may also produce more seeds than small ones.

This will increase the numbers of offspring they contribute to the next **generation** of sunflowers.

Activity 10
Observe Like a Scientist

Quick Code:
ca3385s

Adaptation and Camouflage

Watch the videos about adaptation and camouflage. **Notice** how plants and animals survive in different environments. Then, **talk** about the questions.

Adaptation

Videos

Camouflage

 ## Talk Together

What features or behaviors do plants and animals have that help them survive in their environment?

How might the adaptations of plants and animals provide an advantage when it comes to survival and reproduction?

Activity 11

Evaluate Like a Scientist

Quick Code:
ca3386s

Survival Traits

Examine the images of animals and their specific traits.

Match the trait of each animal to how the trait helps it survive.

CCC **Structure and Function**

Quills of the Porcupine

Color of the Moth

Speed of the Cheetah

Thick Fur of the Polar Bear

Big Teeth of the Beaver

helps it catch its prey

helps it build its home

prevent predators from eating it

helps it hide from predators

helps it stay warm

Activity 12
Evaluate Like a Scientist

Quick Code:
ca3388s

Plant Parent

Bean seeds were allowed to germinate and grow for 30 days.
Their heights are recorded in the table.

Plant	Height of Plant in 30 Days (cm)
A	18
B	4
C	6
D	20
E	28

SEP **Analyzing and Interpreting Data**

Compare the heights of the plant offspring. Circle the correct words or phrases to complete the paragraph.

The offspring of Plant A and Plant D would most likely be
(taller than/shorter than/similar in height to)

the offspring of Plant B and Plant C. The height of the offspring of Plant A and Plant C would most likely be
(taller than/shorter than/similar in height to)

the offspring of Plant B and Plant D. The offspring of Plant A and Plant B would most likely be
(taller than/shorter than)

the offspring of Plant D and Plant E.

Activity 13
Record Evidence Like a Scientist

Quick Code:
ca3389s

Parents and Their Young

Now that you have learned about inherited traits, look again at Parents and Their Young. You first saw this in Wonder.

Let's Investigate Parents and Their Young

Talk Together

How can you describe Parents and Their Young now?
How is your explanation different from before?

SEP **Constructing Explanations and Designing Solutions**

Look at the Can You Explain? question. You first read this question at the beginning of the lesson.

 ## Can You Explain?

How do the traits offspring inherit help them survive?

Now, you will use your new ideas about Parents and Their Young to answer a question.

Write a question. You can use the Can You Explain? question or one of your own. You can also use one of the questions that you wrote at the beginning of the lesson.

My Question

Then, **complete** the chart to answer your question.

Claim: _____

Reasoning	Evidence

S T E M in Action

Activity 14

Analyze Like a Scientist

Careers and Inheritance

Read the text about how people use inheritance to modify other organisms. **Underline** examples of how humans can modify traits in organisms.

Careers and Inheritance

Even though there are many breeds of dog, they are all the same species. How is it possible that a single species can have such varied traits? Dog breeders are people who carefully select traits that are desirable in a dog. They use their knowledge of inheritance to select for the desired trait.

Video

Artificial Selection

Let's take a closer look at how humans have influenced the many different traits we see in various dog breeds. Humans have artificially selected for the dog breeds we see today by observing traits. Large dogs were created by mating a large mother dog with a large father dog. Dog breeders continue to apply **artificial selection** to produce new breeds.

Artificial selection can also be used with plants, including broccoli and strawberries. Botanists, like dog breeders, observe desired traits. Botanists study plants and the way in which they reproduce. Botanists have helped develop plants that can survive even when there is little water. One of these plants is corn. Corn is a grain used to make lots of the food you eat, like cereal.

Video

Engineering and Growing Corn

What Is a Rat Terrier?

Research a breed of dog known as rat terriers. **Find** and **interpret** text resources and images during your research. Why are they named "rat terriers"? What traits do they have that are helpful to humans? Why were these traits selected by dog breeders? **Write** your answers in the space provided.

_____

_____

_____

_____

_____

_____

_____

_____

_____

Artificial Selection

Now it is your turn to think like a botanist! Plant traits can be artificially selected too. **Research** one crop other than corn that has been modified to have desired traits. A few examples of crops are tomatoes, rice, and wheat. But there are many more to choose from. While you research, see if you can think of a reason why artificial selection might be harmful. **Write** your answers in the space provided.

Your answer should include:

1. Name of the crop

2. Description of at least two desired traits

_____

_____

_____

_____

_____

_____

_____

_____

SEP **Obtaining, Evaluating, and Communicating Information**

Activity 15

Evaluate Like a Scientist

Review: Inherited Traits

Quick Code:
ca3391s

Think about what you have read and seen. What did you learn?

Write down some key ideas you have learned. **Review** your notes with a partner. Your teacher may also have you take a practice test.

_____

_____

_____

_____

_____

 Talk Together

Think about the honeybee population loss you read about in Get Started. Use your new ideas about inherited traits to discuss what might be causing honeybee loss.

SEP **Obtaining, Evaluating, and Communicating Information**

© Discovery Education | www.discoveryeducation.com • Image: Simon Eeman / EyeEm / Getty Images

CONCEPT
2.3

Working in Groups

Student Objectives

By the end of this lesson:

☐ I can use evidence to argue that participating in groups helps animals survive.

☐ I can identify limitations of a model ecosystem.

☐ I can describe the effects of changes in a model ecosystem.

☐ I can reason to explain why and how animals work together.

Key Vocabulary

☐ adapt

☐ ecosystem

☐ environment

☐ habitat

☐ migrate

☐ parasite

☐ predator

☐ prey

Quick Code:
ca3393s

Activity 1

Can You Explain?

Why do some animals form groups?

_____

_____

_____

_____

_____

_____

_____

_____

_____

_____

Quick Code:
ca3394s

Activity 2
Ask Questions Like a Scientist

Quick Code:
ca3395s

Whales in Groups

Look at the picture. Then, **answer** the questions.

Let's Investigate Whales in Groups

© Discovery Education | www.discoveryeducation.com • Image: (a) Simon Eeman / EyeEm / Getty Images. (b) Kerstin Meyer / Moment Open / Getty Images

SEP   **Asking Question and Defining Problems**

What do you see? What does this image make you wonder?

_____

_____

_____

_____

_____

How could we answer some of your I wonder statements by
using our observation skills on the image? Where else could
we look to find answers to your questions?

_____

_____

_____

_____

Activity 3

Observe Like a Scientist

© Discovery Education | www.discoveryeducation.com • Image: Simon Eeman / EyeEm / Getty Images

Needs of Living Things

Complete the interactive. **Think** about what each organism needs to survive. Then, **answer** the questions.

Quick Code:
ca3396s

What do living things need to survive?

_____

_____

_____

How do living things meet these needs?

_____

_____

_____

Activity 4

Evaluate Like a Scientist

What Do You Already Know About Working in Groups?

Classifying Needs

Circle the basic needs of most animals.

water

sunlight

food

space

shelter

soil

air

Complete a T-Chart to explain how working in groups helps animals meet their basic needs. In the left column, **list** animal needs. In the right column, **explain** how working in groups helps animals meet each need.

Topic: _____

Animal Needs	Benefit of Forming Groups

Working in Groups

Circle all the sentences that explain why some animals work in groups.

A. They obtain more food for each individual animal compared to the same type of animal looking for food individually.

B. They build bigger shelters than each individual animal would build.

C. They have more success in defending themselves than they would by acting alone.

D. They make faster or more effective adjustments to harmful changes in their ecosystem than they would by acting alone.

E. They can compare themselves to one another in a group, but they have no animals to compare themselves to when they are alone.

Why Do Some Animals Live in Large Groups?

Activity 5

Observe Like a Scientist

Bees and Chimpanzees

Watch the videos about bees and chimpanzees. **Summarize** your evidence on the chart.

Quick Code: ca3398s

How Do Bees Make Honey?

Chimpanzees Use Teamwork

CCC **Patterns**

© Discovery Education | www.discoveryeducation.com • Image: (a) Simon Eeman / EyeEm / Getty Images, (b) Pixabay, (c) Robin Nieuwenkamp / Shutterstock.com

	Bees	Chimpanzees
A = Adjective		
E = Emotion		
I = Interesting		
O = Oh!		
U = Outstanding Question		

Activity 6
Think Like a Scientist

Surviving in a Changing Ecosystem

Quick Code:
ca3399s

In this activity, you will **examine** how animal groups adapt to changing environments.

What materials do you need? (per group)

- Construction paper

- Paper, strips—color #1, 1" × 8.5"

- Paper, strips—color #2, 1" × 8.5"

- Stopwatch

What Will You Do?

You are organisms competing for resources in the environment.

1. **Explore** the environment.

2. **Collect** strips of food, water, and shelter in the time given by your teacher.

| SEP | Developing and Using Models |
| CCC | Stability and Change |

Think About the Activity

What happened when some of the food, water, and shelter resources were removed from the ecosystem?

_____

_____

What happened when some of the organisms in your group had special adaptations to meet their needs?

_____

_____

_____

_____

How did being part of a group help you obtain food and survive in a changing environment?

_____

_____

_____

_____

Teamwork

Read the text about siblings. **Highlight** evidence in the text that supports the claim "Life is easier for animals if they stay in groups." Then, **complete** the graphic organizer.

Teamwork

One way for animals to have a better chance of survival is to form groups with other animals of the same kind.

Groups of animals can be small or large. When members of a group cooperate, they can often accomplish things that one individual could not do on its own.

How do bees work together? Bees need to protect their honey and their offspring. By coming together as a group, they are able to defend themselves against **predators**, like bears. One bee sting would not be enough to defeat a bear, but when

SEP Engaging in Argument from Evidence

Honeybees

many bees sting at the same time, it forces the bear to find another way to satisfy its sweet tooth.

Other animals also work in groups to find food. Herds of elephants find and share food and water. Meerkats teach the young of the group how to hunt and to keep a lookout for predators. Groups help species survive both in good times and during changes to the **ecosystem**.

Elephants

Claim:

Evidence I found: Record all of the evidence you gathered from video, reading, interactives, and hands-on investigations.

My claim is true because:

Activity 8
Evaluate Like a Scientist

Working Together

Quick Code:
ca3401s

Connect each image of animals to the reason that best describes why they are working together.

Wolves hunt as a pack.

Prairie dogs sound the alarm when predators are nearby.

Termites build huge mounds for their colonies.

protect their home   find food or water   adjust to their environment

Bees swarm if predators approach their hive.

Elephants lead each other to rivers to drink.

Bison herds can fend off predators if attacked.

SEP Engaging in Argument from Evidence

DISCOVERY EDUCATION

How Do Animals Work Together in Small Groups?

Activity 9

Observe Like a Scientist

Coyotes, Foxes, and Wolves

Quick Code:
ca3402s

Watch the video. **Look** for changes in animal relationships.

Video

Coyotes, Foxes, and Wolves

Talk Together

Now, talk together about the relationships in the video. What would happen if an elk was sick? What would happen if the coyotes were in a larger pack? Could a coyote hunting alone successfully attack its prey?

CCC **Cause and Effect**

Analyze Like a Scientist

Surviving Together

Quick Code:
ca3403s

Read the text and **watch** the videos. **Highlight** examples of animals working in groups. **Think** about examples of people working in groups. Are any of those examples similar?

Surviving Together

Video

Social Spiders

Video

Bison Animal Cam

Animals work in large groups to defend their homes, to find food and water, and to **adapt** to changes in their **environment**. Animals also work in small groups for the same reasons.

Most commonly, animals will gather in small groups to hunt **prey**. Animals work together in small groups and large groups to help them survive. Wolves work as a pack to hunt for food. Bison travel in herds to protect themselves and their young.

Evaluate Like a Scientist

Quick Code:
ca3404s

Gorilla Groups

The passage tells about life in a gorilla group. Imagine that you need to write a paragraph to argue that group behaviors help young gorillas survive. Which sentences support your argument? **Underline** all the sentences that provide evidence that group behaviors help young gorillas survive.

Gorillas often live in groups. A group is usually made up of a silverback male, a number of females, and their offspring. The silverback male's job is to defend the members of his group from predators and other males. He may weigh nearly 200 kilograms (400 pounds), and he has silver hairs on his back that give him his name. Females usually weigh less than half of what the silverback weighs. The females care for infant gorillas for the first six months of their lives, and infants are never more than five meters away from their mothers for the first year of life.

SEP **Analyzing and Interpreting Data**

Gorilla Groups cont'd

The silverback also keeps the peace, making sure that younger gorillas are not bullied by older ones. When young males become adults, they leave the group and join another group or create their own. Gorillas use sounds like grunts, hoots, and barks to warn one another of predators and to keep themselves from getting lost in the forest. A number of gorilla species live in Africa.

Silverback Gorilla

Activity 12

Record Evidence Like a Scientist

Quick Code:
ca3405s

Whales in Groups

Now that you have learned about working in groups, look again at Whales in Groups. You first saw this in Wonder.

Let's Investigate Whales in Groups

Talk Together

How can you describe Whales in Groups now? How is your explanation different from before?

SEP   **Constructing Explanations and Designing Solutions**

Look at the Can You Explain? question. You first read this question at the beginning of the lesson.

 ## Can You Explain?

Why do some animals form groups?

Now, you will use your new ideas about Whales in Groups to answer a question.

Write a question. You can use the Can You Explain? question or one of your own. You can also use one of the questions that you wrote at the beginning of the lesson.

My Question

Then, **complete** the chart to answer your question.

Claim: _____

Reasoning	Evidence

STEM in Action

Quick Code:
ca3406s

Activity 13

Analyze Like a Scientist

Little Brown Bat

Read the text about bats, and watch the video. **Underline** examples of how living in groups affects the bat.

Little Brown Bat

Bats are an important part of many ecosystems. Many bat species eat primarily insects. Often, these insects are among those that humans consider pests, such as mosquitoes. However, bat populations are suffering because of a disease called white-nose syndrome. This disease is caused by a fungus that grows well in cool, damp places, such as the caves where bats live. It spreads quickly among bat communities because bats live in large groups and come into frequent contact with one another in their caves.

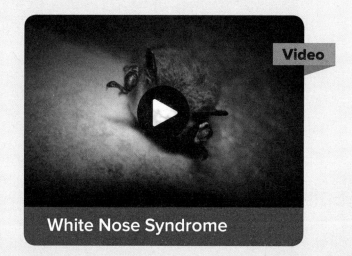

Video

White Nose Syndrome

Wildlife ecologists are people who study animals in their natural environment. These scientists are very concerned about bat species in North America. White-nose syndrome threatens to wipe out several species of bats, some of which are already endangered. One of the great dangers of white-nose syndrome is that the fungus that causes it spreads easily within bat communities. This happens because large groups of bats live together in caves. They huddle together in groups, sharing body heat and communicating with one another. This frequent contact among large numbers of bats means one infected bat can spread the fungus to the entire colony very quickly.

Bats in a Cave

One solution wildlife ecologists suggest is for people to build bat boxes. Bat boxes are similar to birdhouses. People who live in areas where bats are found can build these boxes to house bats in their backyard. The boxes are smaller than caves so not as many bats can fit in them. This means that there are not large groups of bats touching one another. Without regular contact, bats cannot spread the fungus that causes white-nose syndrome as easily.

White-Nose Syndrome Areas

Study this map of the occurrence of white-nose syndrome by county.

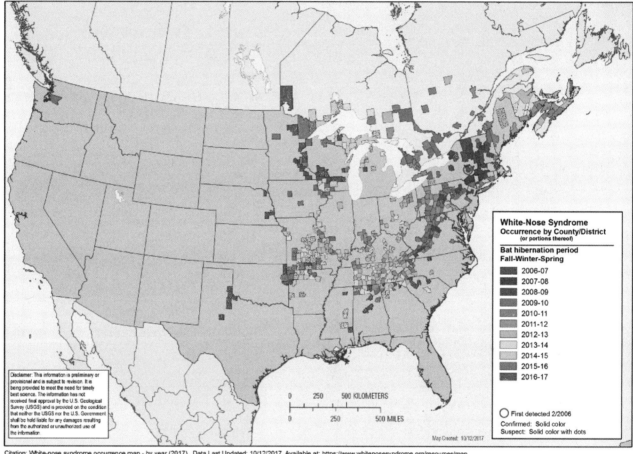

White-Nose Syndrome
Occurrence by County/District
(or portions thereof)

Bat hibernation period
Fall-Winter-Spring

- 2006-07
- 2007-08
- 2008-09
- 2009-10
- 2010-11
- 2011-12
- 2012-13
- 2013-14
- 2014-15
- 2015-16
- 2016-17

○ First detected 2/2006
Confirmed: Solid color
Suspect: Solid color with dots

0 250 500 KILOMETERS
0 250 500 MILES

Disclaimer: This information is preliminary or provisional and is subject to revision. It is being provided to meet the need for timely best science. The information has not received final approval by the U.S. Geological Survey (USGS) and is provided on the condition that neither the USGS nor the U.S. Government shall be held liable for any damages resulting from the authorized or unauthorized use of the information.

Map Created: 10/12/2017

Citation: White-nose syndrome occurrence map - by year (2017). Data Last Updated: 10/12/2017. Available at: https://www.whitenosesyndrome.org/resources/map.

SEP **Using Mathematics and Computational Thinking**

Write a paragraph explaining the threat of white-nose syndrome to bats in your area. **Describe** if the threat is immediate, near future, or distant future. **Give** evidence for your statement and describe your reasoning.

_____

_____

_____

_____

_____

_____

_____

_____

_____

_____

_____

Build a Bat Box

Design a bat box to be placed near your school. **Research** bat boxes online for ideas. **Draw** a diagram of your bat box and label all parts. **Include** a materials list and a list of tools you will need to build the box.

Activity 14

Evaluate Like a Scientist

Review: Working in Groups

Think about what you have read and seen. What did you learn?

Write down some key ideas you have learned. **Review** your notes with a partner. Your teacher may also have you take a practice test.

_____

_____

_____

_____

_____

_____

_____

_____

_____

_____

SEP **Obtaining, Evaluating, and Communicating Information**

DISCOVERY EDUCATION

Talk Together

Think about the honeybee population loss you read about in Get Started. Use your new ideas about animals working in groups to discuss honeybee loss.

Solve Problems Like a Scientist

Quick Code:
ca3418s

Unit Project: Honeybee Population Loss

In this project, you will use what you know about how animals work together to help keep bees alive.

In recent years, scientists, farmers, and others have become very concerned about honeybee populations. Bees play an important role in ecosystems. They act as pollinators, which means they move pollen from plant to plant. This allows plants to reproduce. But bees are dying. Entire bee colonies die every year.

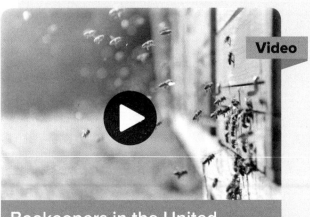

Beekeepers in the United States Lost 44 Percent of Their Hives over 12 Months

Scientists at the United States Department of Agriculture (USDA) investigate bee colony losses. They survey the number of hives during the year. Then, they calculate the change in the number of hives

Honeybees on a Comb

from the previous year. Over the past several years, they have seen that large numbers of hives have died. The main cause of death is the varroa mite, a parasite that attacks bee colonies. Other causes of death include other parasites, diseases, and pesticides.

Graphing Bee Losses

Look at the data in the table. The data shows the number of bee colonies lost each quarter from the year 2015 to 2017. **Use** the data to draw a a bar graph.

2015 Q1	2015 Q2	2015 Q3	2015 Q4	2016 Q1
92,250	56,760	88,390	87,610	115,950
2016 Q2	**2016 Q3**	**2016 Q4**	**2017 Q1**	**2017 Q2**
47,780	92,610	129,290	84,430	34,750

SEP **Asking Questions and Defining Problems**

Bee Colonies Lost

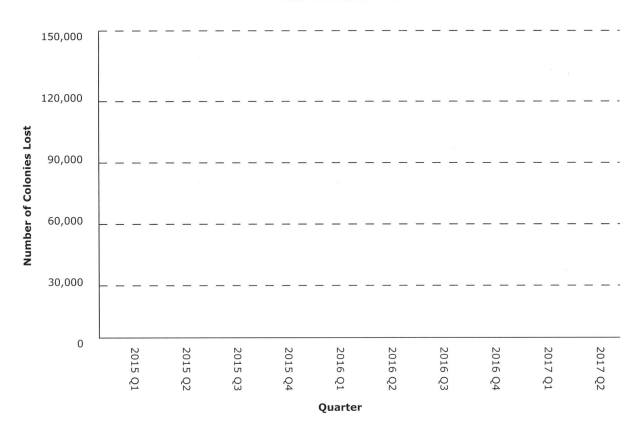

Number of Colonies Lost (y-axis): 0, 30,000, 60,000, 90,000, 120,000, 150,000

Quarter (x-axis): 2015 Q1, 2015 Q2, 2015 Q3, 2015 Q4, 2016 Q1, 2016 Q2, 2016 Q3, 2016 Q4, 2017 Q1, 2017 Q2

Bee Loss Control

Conduct online research about the ways people help to reduce honeybee colony loss. **Write** a paragraph describing these methods. **Describe** at least three methods and include an analysis of how each method reduces bee deaths.

_____

_____

_____

_____

_____

_____

_____

_____

_____

_____

_____

_____

_____

_____

_____

_____

_____

_____

_____

Now **discuss** your findings with your class. **Choose** the
method you prefer most. **Explain** why you prefer that method.

Grade 3 Resources

- **Bubble Map**
- **Safety in the Science Classroom**
- **Vocabulary Flash Cards**
- **Glossary**
- **Index**

Bubble Map

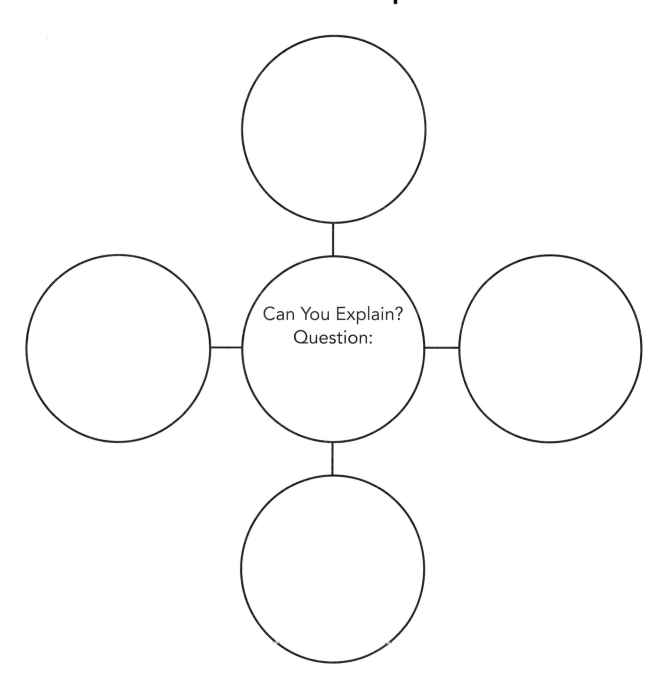

Can You Explain?
Question:

Safety in the Science Classroom

Following common safety practices is the first rule of any laboratory or field scientific investigation.

Dress for Safety
One of the most important steps in a safe investigation is dressing appropriately.

- Splash goggles need to be kept on during the entire investigation.

- Use gloves to protect your hands when handling chemicals or organisms.

- Tie back long hair to prevent it from coming in contact with chemicals or a heat source.

- Wear proper clothing and clothing protection. Roll up long sleeves, and if they are available, wear a lab coat or apron over your clothes. Always wear close toed shoes. During field investigations, wear long pants and long sleeves.

Be Prepared for Accidents
Even if you are practicing safe behavior during an investigation, accidents can happen. Learn the emergency equipment location in your classroom and how to use it.

- The eye and face wash station can help if a harmful substance or foreign object gets into your eyes or onto your face.

- Fire blankets and fire extinguishers can be used to smother and put out fires in the laboratory. Talk to your teacher about fire safety in the lab. He or she may not want you to directly handle the fire blanket and fire extinguisher. However, you should still know where these items are in case the teacher asks you to retrieve them.

- Most importantly, when an accident occurs, immediately alert your teacher and classmates. Do not try to keep the accident a secret or respond to it by yourself. Your teacher and classmates can help you.

Practice Safe Behavior

There are many ways to stay safe during a scientific investigation. You should always use safe and appropriate behavior before, during, and after your investigation.

Safety Goggles

- Read the all of the steps of the procedure before beginning your investigation. Make sure you understand all the steps. Ask your teacher for help if you do not understand any part of the procedure.

- Gather all your materials and keep your workstation neat and organized. Label any chemicals you are using.

- During the investigation, be sure to follow the steps of the procedure exactly. Use only directions and materials that have been approved by your teacher.

- Eating and drinking are not allowed during an investigation. If asked to observe the odor of a substance, do so using the correct procedure known as wafting, in which you cup your hand over the container holding the substance and gently wave enough air toward your face to make sense of the smell.

- When performing investigations, stay focused on the steps of the procedure and your behavior during the investigation. During investigations, there are many materials and equipment that can cause injuries.

- Treat animals and plants with respect during an investigation.

- After the investigation is over, appropriately dispose of any chemicals or other materials that you have used. Ask your teacher if you are unsure of how to dispose of anything.

- Make sure that you have returned any extra materials and pieces of equipment to the correct storage space.

- Leave your workstation clean and neat. Wash your hands thoroughly.

adapt

Image: Paul Fuqua

something a plant or animal does to help it survive in its environment

adaptation

Image: Paul Fuqua

how a plant or animal has changed over time to help it survive in its environment

artificial selection

Image: DanielBrachlow/Pixabay

specifically breeding animals or cultivating plants only for certain desired genetic outcomes

camouflage

Image: Discovery Communications, Inc.

the coloring or patterns on an animal's body that allow it to blend in with its environment

characteristic

Image: Discovery Communications, Inc.

a feature of an organism; something you can observe about an organism

cycle

Image: inbevel / Shutterstock.com

a process that repeats

ecosystem

Image: Paul Fuqua

all the living and nonliving things in an area that interact with each other

endangered

Image: Paul Fuqua

a species at risk of becoming extinct

environment

Image: Odua Images / Shutterstock.com

all the living and nonliving things that surround an organism

extinct

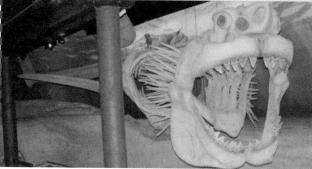

Image: Ocean First Education

when a species is no longer surviving

generation

Image: Paul Fuqua

the next group of living things or species that will be born around the same time

germination

Image: Paul Fuqua

the process in which a young plant sprouts from a seed

habitat

Image: Paul Fuqua

the location in which an organism lives

inherit

Image: Discovery Communications, Inc.

to receive genetic information and traits from a parent or parents

life cycle

Image: Paul Fuqua

the various stages of an organism's development and reproduction

lifespan

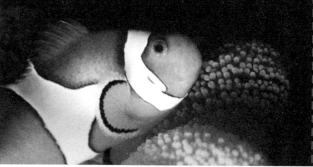

Image: Paul Fuqua

how long in time an organism is expected to live

mature

Image: Paul Fuqua

(adj) describes an organism that is fully grown or adult; (v)to grow up; to become an adult

metamorphosis

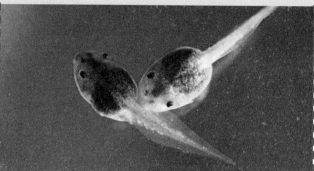

Image: Paul Fuqua

process in which an animal's body undergoes dramatic changes in form during its life cycle

migrating

Image: Discovery Communications, Inc.

traveling within a group to a different location during season changes

offspring

Image: Paul Fuqua

a new organism that is produced by one or more parents

organism

Image: Discovery Communications, Inc.

any individual living thing

parasite

Image: Pixabay

something that lives on or in another living thing that provides its food

pollen

Image: Paul Fuqua

a yellow powder produced by the stamen of a flower; pollen fertilizes the pistil of another flower

predator

Image: Ocean First Education

the larger animals that hunt the smaller animals, or prey, for food

prey

Image: Discovery Communications, Inc.

the animals that get hunted by the larger animals, or predators, for food

seedling

Image: Pixabay

a young plant that grows from a seed

seeds

Image: Paul Fuqua

the part of a plant that contains a young plant, food supply, and protective coating

trait

Image: Paul Fuqua

a characteristic or property of an organism

English ——— A ——— Español

adapt

something a plant or animal does to help it survive in its environment

adaptarse

algo que una planta o animal hace para sobrevivir en su medio ambiente

adaptation

how a plant or animal has changed over time to help it survive in its environment (related word: adapt)

adaptación

proceso mediante el cual las características de una especie cambian a través de varias generaciones como respuesta al medio ambiente (palabra relacionada: adaptar)

adjust

to change one's position or behavior to allow for a better fit, to adapt

acomodarse

cambiar de posición o comportamiento para ajustarse mejor, adaptarse

air

a gas that is all around you and you can't see it, but living things like plants and animals need it to breathe and to grow

aire

parte de la atmósfera más cercana a la Tierra; la parte de la atmósfera que los organismos que habitan la Tierra utilizan para respirar

air pressure

the force that air puts on an area (related word: pressure)

presión de aire

fuerza que el aire ejerce sobre un área (palabra relacionada: presión)

---

analyze

to closely examine something and then explain it

analizar

examinar con atención algo y luego explicarlo

---

ancient

very old

antiguo

extremadamente viejo

---

arctic

being from an icy climate, such as the North Pole

ártico

que tiene relación con el Polo Norte o el área que lo rodea

---

artificial selection

specifically breeding animals or cultivating plants only for certain desired genetic outcomes

selección artificial

específicamente criar animales o cultivar plantas sólo para determinados resultados genéticos deseados

atmosphere

layers of gas that surround a planet (related word: atmospheric)

atmósfera

capas de gas que rodean un planeta (palabra relacionada: atmosférico)

attract

to pull one thing toward another (related word: attraction)

atraer

tirar o acercar hacia un punto; los polos opuestos de un imán se unen cuando se atraen (palabra relacionada: atracción)

B

balanced forces

when two equal forces are applied to an object in opposite directions, the object does not move

fuerza equilibrada

cuando se aplican dos fuerzas iguales sobre un objeto en direcciones opuestas, el objeto no se mueve

barometer

a tool used to measure air pressure (related word: barometric)

barómetro

herramienta usada para medir la presión del aire (palabra relacionada: barométrico)

behavior
the way in which a living thing acts (related word: behave)

conducta
todas las acciones y reacciones de un animal (palabra relacionada: comportarse)

— C —

camouflage
the coloring or patterns on an animal's body that allow it to blend in with its environment

camuflaje
color o apariencia del cuerpo de un animal que le permite mezclarse con su medioambiente

carnivore
a meat eater

carnívoro
que se alimenta de carne

characteristic
a special quality that something may have

característica
rasgo de un organismo; algo que se puede observar sobre un organismo

climate
the usual weather conditions in a place or area (related word: climatic)

clima
condiciones promedio del tiempo en un área (palabra relacionada: climático)

coast
an area where the ocean meets
the land

costa
un área donde el océano se
encuentra con la tierra

community
a group of different populations
that live together and interact in
an environment

comunidad
grupo de distintas poblaciones
que viven juntas e interactúan en
un ambiente

contact
when two things are so close
they touch

contacto
cuando dos objetos están tan
cerca que se tocan

coral reef
an area that forms in the warm,
shallow ocean waters made from
the hard skeletons of animals
called corals

arrecife de coral
estructura formada por un
esqueleto duro de animales que
viven en aguas oceánicas cálidas
y poco profundas

cycle
a process that repeats (related
word: cyclical)

ciclo
proceso que se repite (palabra
relacionada: cíclico)

D

data
measurements or observations
(related word: datum)

datos
medidas u observaciones
(palabra relacionada: dato)

desert
an area that gets very little rain
water and does not have a lot of
growing plants

desierto
área que recibe muy poca
precipitación y tiene muy poca
vegetación

detect
to notice or find, often with the
help of a science tool (related
words: detection, detector)

detectar
notar o encontrar, generalmente
con la ayuda de una herramienta
científica (palabras relacionadas:
detección, detector)

dinosaur
an extinct organism with reptile
and birdlike features: Dinosaurs
lived on Earth millions of years
ago

dinosaurio
organismo en extinción con
características de reptil y ave: los
dinosaurios vivían en la Tierra
hace millones de años

discharge
the release of energy

descarga
liberación de energía

drought
a long period of little or no rain

sequía
falta prolongada de lluvia

ecosystem
all the living and nonliving things in an area that interact with each other

ecosistema
todos los seres vivos y objetos sin vida de un área, que se interrelacionan entre sí

electrical charges
a type of charge, either positive, negative, or neutral

carga eléctrica
un tipo de carga, ya sea positiva, negativa, o neutra

electrical energy
energy produced by power plants that flows through electrical lines and wires

energía eléctrica
energía producida por centrales eléctricas que fluye a través de cables y líneas eléctricas

electromagnet
a metal object that acts as a magnet when an electric current moves through it

electroimán
objeto de metal que actúa como un imán cuando una corriente eléctrica se mueve a través de él

endangered

a type of plant or animal that is in danger of becoming extinct

amenazado

un tipo de planta o animal que está en peligro de extinción

energy

the ability to do work or make something change

energía

habilidad para hacer un trabajo o producir un cambio

environment

all the living and nonliving things that surround an organism

medio ambiente

todos los seres vivos y objetos sin vida que rodean a un organismo

equator

an imaginary line that divides Earth into Northern and Southern Hemispheres; located halfway between the North and South Poles (related word: equatorial)

ecuador

línea imaginaria que divide la Tierra en Hemisferio Norte y Hemisferio Sur; ubicada a mitad de camino entre el Polo Norte y el Polo Sur (palabra relacionada: ecuatorial)

evidence

facts that give us more information, clues, or proof about something else

evidencia

hechos que nos dan más información, pistas o pruebas sobre otra cosa

extinct

when a plant or an animal is no longer in existance (related word: extinction)

extinto

palabra que hace referencia a una especie de animales que una vez habitó la Tierra, pero que ya no existe (palabra relacionada: extinción)

F

---

factor

something that influences another thing to move or change

factor

algo que influye en que otra cosa se mueva o cambie

food web

a model that shows many different feeding relationships among living things

red alimentaria

modelo que muestra muchas y diferentes relaciones de alimentación entre los seres vivos

force

a pull or push that is applied to an object

fuerza

acción de atraer o empujar que se aplica a un objeto

forecast

(v) to analyze weather data and make an educated guess about weather in the future; (n) a prediction about what the weather will be like in the future based on weather data

pronosticar / pronóstico

(v) analizar los datos del tiempo y hacer una conjetura informada sobre el tiempo en el futuro; (s) predicción sobre cómo será el tiempo en el futuro en base a datos

fossil

the remains of a living animal or plant from a very long time ago (related word: fossilize)

fósil

muestra de que un organismo existió una vez en un área; puede ser una parte del cuerpo del organismo o un rastro fósil, que es una marca o impresión dejada por el organismo (palabra relacionada: fosilizar)

friction

when two objects rub against each other

fricción

fuerza que se opone al movimiento de un cuerpo sobre una superficie o a través de un gas o un líquido

G

generation
the next group of living things or species that will be born around the same time

generación
el siguiente grupo de seres vivos o especies que nacerán alrededor de la misma época

germination
when a young plant grows from a seed

germinación
proceso por el cual una planta joven brota de una semilla (palabra relacionada: germinar)

grassland
a large area of land covered by grass

pradera
área de tierra cubierta principalmente de pasto con pocos arbustos o árboles

gravity
the force that pulls an object toward the center of Earth (related word: gravitational)

gravedad
fuerza que existe entre dos objetos cualquiera que tienen masa (palabra relacionada: gravitacional)

habitat
the place where a plant or animal lives

hábitat
lugar donde vive un organismo

heat
a form of energy; the state of being very warm

calor
transferencia de energía térmica

herbivore
a plant eater

herbívoro
que se alimenta de vegetales

humidity
the measure of how much water vapor is in the air

humedad
medida de cuánto vapor de agua hay en el aire

hurricane
a storm with strong winds and rain that forms over tropical waters (related terms: typhoon, tropical cyclone)

huracán
tormenta con fuertes vientos y lluvia que se forma sobre aguas tropicales (palabras relacionadas: tifón, ciclón tropical)

I

impact
to influence or affect something

impactar
afectar o influir en algo

inherit
to get genetic information and traits from a parent or parents (related word: inheritance)

heredar
obtener información y rasgos genéticos de uno o ambos padres (palabra relacionada: herencia)

instinct
behaviors animals and people are born with that help them survive

instinto
conductas con las que nacen los animales y las personas y que los ayudan a sobrevivir

interact
to act on one another (related word: interaction)

interactuar
ejercer influencia mutua (palabra relacionada: interacción)

L

life cycle
the various stages of an organism's development and reproduction

ciclo de la vida
diversas etapas del desarrollo y de la reproducción de un organismo

lifespan

how long in time an organism is expected to live

longevidad

cuánto tiempo se espera que viva un organismo

lightning

when electricity flows between a cloud and the ground or between two clouds and you sometimes see a streak or a flash in the sky

relámpago

descarga grande de electricidad en el aire que equilibra una diferencia en la carga eléctrica que existe entre una nube y otra, o entre las nubes y el suelo

— M —

magnetic

having the properties of a magnet; having the ability to be attracted to or by a magnet

magnético

que tiene las propiedades de un imán; que tiene la capacidad de ser atraído hacia o por un imán

magnetic field

a region in space near a magnet or electric current in which magnetic forces can be detected

campo magnético

región en el espacio cerca de un imán o de una corriente eléctrica, donde pueden detectarse fuerzas magnéticas

magnetism
the amount of attraction to a magnet

magnetismo
la cantidad de atracción hacia un imán

mature
when a living thing is fully grown or an adult (related word: maturity)

maduro / madurar
describe un organismo que ha crecido por completo, o que es adulto (palabra relacionada: madurez)

metamorphosis
when a living thing goes through changes during its life cycle, like a frog or a butterfly

metamorfosis
proceso por el cual el cuerpo de un animal experimenta cambios drásticos en su forma durante su ciclo de vida

meteorology
the study of patterns of weather

meteorología
estudio de los patrones del tiempo

microorganisms
the tiniest of living things, can only be seen under a microscope

microorganismo
los seres vivos más diminutos, que sólo se pueden ver bajo un microscopio

migrating

traveling within a group to a different location during season changes

migratorio

que viaja dentro de un grupo a un lugar diferente durante los cambios de estación

motion

when something moves from one place to another (related terms: move, movement)

movimiento

cambio en la posición de un objeto en comparación con otro objeto (palabras relacionadas: mover, desplazamiento)

N

natural

not human-made (related word: nature)

natural

que no está hecho por un ser humano (palabra relacionada: naturaleza)

negative charge

a charge that you get when there is a build-up of electrons

carga negativa

carga eléctrica en la cual hay una selección de electrones; opuesto a carga positiva

neutral
having no electrical charge, being neither positive nor negative

neutral
los objetos neutros no tienen carga eléctrica

nutrient
something in food that helps people, animals and plants live and grow

nutriente
sustancia como la grasa, una proteína o un carbohidrato que un ser vivo necesita para sobrevivir

O

observe
to study something using your senses (related word: observation)

observar
estudiar algo usando tus sentidos (palabra relacionada: observación)

offspring
a new organism that is produced by one or more parents

descendencia
organismo nuevo que es el producto de la reproducción

organism
any individual living thing

organismo
todo ser vivo individual

parasite

a plant, animal, or fungus that lives on or in another living thing to get food and energy from it

parásito

una planta, animal, u hongo que vive sobre o dentro de otro ser vivo, del cual se alimenta y obtiene energía

pendulum

a string or bar that is loose at one end but fixed at the other end and can swing back and forth, like in a clock

péndulo

cuerda o barra que posee un extremo suelto y otro fijo y puede rotar o balancearse alrededor del punto fijo

pole

the opposite ends of a battery, a magnet, or the north and south ends of Earth

polo

extremos opuestos de una batería eléctrica, un imán, o los extremos norte y sur de Tierra

pollen

the yellow powder found inside of a flower (related word: pollinate)

polen

polvo amarillo producido por el estambre de una flor; el polen fertiliza el pistilo de otra flor (palabra relacionada: polinizar)

pollution

when harmful materials have been put into the air, water, or soil (related word: pollute)

contaminación

cuando se introducen materiales perjudiciales en el aire, el agua o el suelo (palabra relacionada: contaminar)

positive charge

a charge that you get when there are more protons than electrons

carga positiva

Carga electrónica con más protones que neutrones; lo opuesto a una carga negativa

precipitation

water that is released from clouds in the sky; includes rain, snow, sleet, hail, and freezing rain

precipitación

agua liberada de las nubes en el cielo; incluye la lluvia, la nieve, la aguanieve, el granizo, y la lluvia congelada

predators

the larger animals that hunt the smaller animals, or prey, for food

depredador

animal que caza y come a otro animal

predict

to make a guess based on what you already know (related word: prediction)

predecir

adivinar qué sucederá en el futuro (palabra relacionada: predicción)

prehistoric

a time before history was written

prehistórico

de una época anterior a la que registrara la historia de la humanidad

prey

the animals that get hunted by the larger animals, or predators, for food

presa

animal que es cazado y comido por otro

─────── **R** ───────

rain

liquid water that falls from the sky

lluvia

agua líquida que cae desde el cielo

recycle

to create new materials from used products

reciclar

crear nuevos materiales a partir de productos usados

region

a place, especially around the world

región

lugar, especialmente alrededor del mundo

repel
to force an object away or to keep it away

repeler
forzar a un objeto para que se aleje o mantenerlo alejado

reproduce
to make more of a species; to have offspring (related word: reproduction)

reproducir
hacer más de una especie; tener descendencia (palabra relacionada: reproducción)

— **S** —

seed
the small part of a flowering plant that grows into a new plant

semilla
parte pequeña de una planta con flor que crece y se convierte en una nueva planta

seedling
a young plant that grows from a seed

plántula
planta joven que crece de una semilla

severe
dangerous or harsh conditions

severo
condiciones peligrosas o duras

species
a group of the same kinds of living things

especie
un grupo de las mismas clases de seres vivos

static electricity
electric charges that build up on an object

electricidad estática
cargas eléctricas que se acumulan sobre un objeto

stored energy
energy in an object or substance that is not being given off by the object or substance

energía almacenada
energía en un objeto o una sustancia que no es liberada por el objeto o la sustancia

survival
ability to live and remain alive

supervivencia
capacidad de vivir y mantenerse vivo

survive
to continue living or existing: an organism survives until it dies; a species survives until it becomes extinct (related word: survival)

sobrevivir
continuar viviendo o existiendo: un organismo sobrevive hasta que muere; una especie sobrevive hasta que se extingue (palabra relacionada: supervivencia)

T

temperature (general)
a measure of how hot or cold a
substance is

temperatura (general)
medida de cuán caliente o fría es
una sustancia

tornado
a funnel-shaped cloud or column
of air that rotates at high speeds
and extends downward from a
cloud to the ground

tornado
nube o columna de aire con
forma de embudo que rota a
altas velocidades y se extiende
hacia abajo desde una nube al
suelo

trait
a characteristic that you get from
one of your parents

rasgo
característica o propiedad de un
organismo

tropical
from a warmer climate, near the
equator

tropical
de un clima más cálido, cerca del
ecuador

W

water
a clear liquid that has no taste or
smell

agua
compuesto formado por
hidrógeno y oxígeno

weather

the properties of the atmosphere at a given time and location, including temperature, air movement, and precipitation

tiempo atmosférico

propiedades de la atmósfera en un determinado momento y lugar; entre ellas, la temperatura, el movimiento, de aire y las precipitaciones

wind

the movement of air due to atmospheric pressure differences

viento

movimiento de aire que se produce por las diferencias en la presión atmosférica

work

a force applied to an object over a distance

trabajo

fuerza aplicada a un objeto a lo largo de una distancia

Index

A

Adapt 110

Adaptation 58, 77

Analyze Like a Scientist 12–15, 24–25, 33–34, 43–47, 66–67, 70–71, 74–76, 85–88, 104–107, 110, 117–121

Artificial selection 86

Ask Questions Like a Scientist 10–11, 54–55, 94–95

C

Camouflage 77

Can You Explain? 8, 41, 52, 83, 92, 115

Characteristics

    as survival 75

    inherited 71

    or traits 66–67

Cycle, life

    interruption of 43–45

    models of 30

    of animals 24–25

    of plants 33–34

E

Ecosystems

    changing 124–125

    diseases in 117–118

    groups in 105

Endangered 43–45

Environment 110

Evaluate Like a Scientist 17–19, 30–31, 35, 38, 48–49, 56–59, 72–73, 78–79, 80–81, 89, 97–99, 108, 111–112, 122–123

Extinct 43

G

Generation 76

Germination 33

H

Habitat 44

Hands-On Activities 20–23, 60–65

I

Inherit 71, 85

Investigate Like a Scientist 20–23, 60–65

L

Life cycle

    interruption of 43–45

    models of 30

    of animals 24–25

    of plants 33–34

Lifespan 13

M

Mature 33

Metamorphosis 25, 30–31

O

Observe Like a Scientist 16, 26–27, 28–29, 32, 36–37, 68–69, 77, 96, 100–101, 109

Offspring

of animals 24–25

parents and 25, 71

Organisms

adaptations and traits of 58, 85

as part of a life cycle 13–15, 24–25

P

Parasite 125

Plants

by artificial selection 85–86

life cycle of 33–34

Pollen 124

Predator 104–105, 112

Prey 110

R

Record Evidence Like a Scientist 40–42, 82-84, 114–116

S

Seeds

in the plant life cycle 33–34

inherited traits 75

Seedlings 33–35

STEM in Action 43–47, 85–88, 117–121

T

Think Like a Scientist 102–103

Traits

and survival 74–75

inherited 66–67, 71, 85–88

U

Unit Project 4–5, 124–127